ISA-S77.41
Approved November, 1992

Standard

Fossil Fuel Power Plant Boiler Combustion Controls

Instrument Society of America

ISBN 1-55617-108-0

ISA-S77.41, Fossil Fuel Power Plant Boiler Combustion Controls

INSTRUMENT SOCIETY OF AMERICA
67 Alexander Drive
Research Triangle Park, North Carolina 27709

Preface

This preface is included for informational purposes and is not a part of ISA-S77.41.

This standard has been prepared as a part of the service of the Instrument Society of America (ISA) towards a goal of uniformity in the field of instrumentation. To be of real value, this document should not be static, but should be subject to periodic review. Toward this end, the Society welcomes all comments and criticisms and asks that they be addressed to the Secretary, Standards and Practices Board, Instrument Society of America, 67 Alexander Drive, P.O. Box 12277, Research Triangle Park, North Carolina 27709, Telephone (919) 549-8411.

The ISA Standards and Practices Department is aware of the growing need for attention to the metric system of units in general, and the International System of Units (SI) in particular, in the preparation of instrumentation standards. The Department is further aware of the benefits to U. S. A. users of ISA Standards of incorporating suitable references to the SI (and the metric system) in their business and professional dealings with other countries. Toward this end, this Department will endeavor to introduce SI-acceptable metric units in all new and revised Standards to the greatest extent possible. Metric Practice, which has been published by the Institute of Electrical and Electronic Engineers as ANSI/IEEE Standard 268-1982, and future revisions will be reference guide for definitions, symbols, abbreviations, and conversion factors.

It is the policy of the Instrument Society of America to encourage and welcome the participation of all concerned individuals and interests in the development of ISA Standards. Participation in the ISA standards-making process by an individual in no way constitutes endorsement by the employer of that individual, of the Instrument Society of America, or any of the Standards that ISA develops.

The information contained in this preface, the footnotes, and appendices is included for information only and it not part of the standard.

The following people served as members of ISA Subcommittee SP77.41:

Name	Company
G. McFarland, Chairman	ABB Power Plant Controls
H. Hopkins, Managing Director	Utility Products of Arizona
E. Adamson	Stone & Webster, Inc.
M. Blaschke	Weyerhaeuser Company
W. Blazier	Sargent & Lundy
L. Broeker	PSI Energy
N. Burris	Tennessee Valley Authority
R. Campbell	Salt River Project
* T. Dimmery	Duke Power Company
R. Eng	Foster Wheeler Energy Corporation
D. Frey	Computer & Control Consultant
M. Fryman	Illinois Power Company
W. Holland	Southern Company Services,Inc.
H. Hubbard	Retired
R. Hubby	Leeds & Northrup
L. Hughart	Union Carbide Company
* M. Laney	Duke Power Company
W. Matz	Forney, Int.

Name	Company
E. McWilliams	Consultant
J. Murphy	E. I. du Pont
T. Russell	Honeywell, Inc.
A. Schager	Consultant
R. Spellman	Westinghouse Electric Corporation
O. Taneja	Consultant
T. Toms	Carolina Power & Light Company
B. Traylor	Sandwell, Inc.
* H. Wall	Duke Power Company

*One Vote

The following people served as members of ISA Committee SP77:

Name	Company
R. Johnson, Chairman	Sargent & Lundy
* T. Russell, Co-Chairman	Honeywell, Inc.
H. Hopkins, Managing Director	Utility Products of Arizona
M. Fryman, Secretary	Illinois Power Company
L. Altcheh	Israel Electric Corporation
W. Blazier	Sargent & Lundy
L. Broeker	PSI Energy
J. Cartwright	Ohmart Corporation
Q. Chou	Ontario Hydro
* T. Dimmery	Duke Power Company
C. Fernandez	Comision Federal de Electricidad
D. Foreman	Brown & Root, Inc .
A. Gile	Potomac Electric Company
R. Hicks	Black & Veatch
* W. Holland	Southern Company Services, Inc.
* R. Hubby	Leeds & Northrup
R. Karvinen	EG & G Idaho, Inc.
D. Lee	Bailey Controls, Company
W. Matz	Forney, International
G. McFarland	ABB Power Plant Controls
J. Moskal	ABB Combustion Engineer Systems
* J. Murphy	E. I. du Pont
* T. New	Leeds & Northrup
* N. Obleton	Honeywell, Inc.
R. Papilla	Consultant
* J. Pickle	E. I. du Pont
R. Ramachandran	American Cyanamid Company
D. Rawlings, II	Babcock & Wilcox, Company
R. Roop	Hooser Energy, Inc.
J. Rutledge	Jacksonville Electric Authority
A. Schager	Consultant
J. Smith	Westinghouse Electric Corporation
T. Stevenson	Baltimore Gas & Electric Company

Name	Company
* C. Taft	Southern Company Services, Inc.
D. Tennant	Georgia Power Company
B. Traylor	Sandwell, Inc.
* H. Wall	Duke Power Company
J. Weiss	EPRI
J. Wiley	Public Service of Colorado

This draft standard was approved for publication by the ISA Standards and Practices Board in November 1992.

Name	Company
W. Weidman, Vice President	Gilbert Commonwealth, Inc.
M. Widmeyer, Vice President Elect	The Supply System
H. Baumann	H.D. Baumann & Associates Ltd.
C. Gross	Dow Chemical Company
H. Hopkins	Utility Products of Arizona
A. Iverson	Lyondell Petrochemical Company
K. Lindner	Endress & Hauser GmbH & Company
G. McFarland	ABB Power Plant Controls
E. Nesvig	ERDCO Engineering Corporation
D. Rapley	Rapley Engineering Services
R. Reimer	Allen-Bradley Company
J. Rennie	Factory Mutual Research Corporation
R. Webb	Pacific Gas & Electric Company
J. Weiss	Electric Power Research Institute
J. Whetstone	National Institute of Standards & Tech.
C. Williams	Eastman Kodak Company
M. Zielinski	Rosemount, Inc.
* D. Bishop	Chevron U.S.A., Inc.
* P. Bliss	Consultant
* W. Calder, III	Consultant
* B. Christensen	Consultant
* L. Combs	Retired/Consultant
* N. Conger	Consultant
* T. Harrison	FAMU/FSU College of Engineering
* R. Jones	Retired/Consultant
* R. Keller	Engineering Support Services
* O. Lovett, Jr.	Consultant
* E. Magison	Honeywell, Inc.
* R. Marvin	Roy G. Marvin Company
* A. McCauley, Jr.	Chagrin Valley Controls, Inc.
* W. Miller	Retired/Consultant
* J. Mock	Bechtel
* G. Platt	Retired/Consultant
* R. Prescott	Moore Products Company
* C. Reimann	National Institute of Standards & Tech.
* K. Whitman	ABB Combustion Engineering
* J. Williams	Consultant

* Directors Emeriti

Foreword

A variety of combustion control systems have been developed over the years to fit the needs of particular applications. Load demand, operating philosophy, plant layout, and type of firing must be considered before the ultimate selection of a system is made. Therefore, this standard is not intended to limit the complexity or scope of the combustion control system design that one might wish to implement, but rather to establish a minimum of control needed.

This standard is part of a series resulting from the efforts of the SP77 Committee on Fossil Power Plant Standards, especially subcommittee SP77.40 on Boiler Controls. It should be used in conjunction with the other SP77 series of standards for safe, reliable, and efficient design, construction, operation, and maintenance of the power plant. It is not the intent of this standard to establish any procedures or practices that are contrary to any other standard in this series.

Table of Contents

1 Purpose

1.1 The purpose of this standard is to establish the minimum requirements for the functional design specification of combustion control systems for drum-type fossil-fueled power plant boilers.

2 Scope

2.1 The scope of this standard is to address the major combustion control subsystems in boilers with steaming capabilities of 200,000 lb/hr (25 kg/s) or greater. These subsystems include, but are not limited to, furnace pressure control (balanced draft), air flow control, and fuel flow control when firing coal, oil, gas, or combinations thereof. Specifically excluded from consideration are development of boiler energy demand, all burner control, interface logic systems, and associated safety systems, as well as all controls associated with fluidized bed and stoker-fired combustion units.

3 Definitions

The following definitions are provided to clarify their use in this standard and may not be relevant to the use of the word in other texts. For other definitions, please refer to ISA-S51.1, Process Instrumentation Terminology.

air: The mixture of oxygen, nitrogen, and other gases, which, with varying amounts of water vapor, forms the atmosphere of the earth.

air purge: A flow of air through the furnace, boiler gas passages, and associated flues and ducts that will effectively remove any gaseous combustibles and replace them with air. Purging may also be accomplished by an inert medium.

balanced draft: A system of furnace pressure control in which the inlet air flow or the outlet flue gas flow is controlled to maintain the furnace pressure at a fixed value (typically slightly below atmospheric).

boiler: A closed vessel in which water is heated, steam is generated, steam is superheated, or any combination thereof, by the application of heat from combustible fuels in a self-contained or attached furnace.

combustion: The rapid chemical combination of oxygen with the combustible elements of a fuel, resulting in the production of heat.

combustible: The heat producing constituent of a fuel, flue gas, or fly ash.

control (controller): Any manual or automatic device or system of devices used to regulate processes within defined parameters.

convection: The transmission of heat by the circulation of a liquid or a gas such as air. Convection may be natural or forced.

draft: The difference between atmospheric pressure and some lower pressure existing in the furnace or gas passages of a steam-generating unit.

efficiency: The ratio of energy output to the energy input. The efficiency of a boiler is the ratio of heat absorbed by water and steam to the heat equivalent of the fuel fired.

excess air: Air supplied for combustion in excess of theoretical combustion air.

flue gas: The gaseous products of combustion in the flue to the stack.

forced draft fan: A fan supplying air under pressure to the fuel-burning equipment.

fuel: A substance containing combustible material used for generating heat; coal, oil, and gas are fuels referenced in this standard.

furnace: An enclosed space provided for the combustion of fuel.

furnace pressure: The pressure of gases in the furnace (see also draft).

induced draft fan: A fan exhausting flue gases from the furnace.

load: The rate of output.

primary air: (transport air, pulverizer air): the air or flue gas introduced into the pulverizer to dry the fuel and convey the pulverized fuel to the burners.

redundant (redundance): Duplication or repetition of elements in electronic or mechanical equipment to provide alternative functional channels in case of failure.

secondary air: The air supplied by the forced draft fan to the burners for combustion.

secondary combustion: Combustion that occurs as a result of ignition at a point beyond the furnace.

shall, should*: The word "shall" is to be understood as a REQUIREMENT; the word "should" as a RECOMMENDATION.

tertiary air: The air supplied to certain types of burners for cooling the burner metal or to improve the combustion process.

theoretical (stoichiometric) combustion air: The chemically correct amount of air required for complete combustion of a given quantity of a specific fuel.

**Per ANSI Style Manual.*

4 Minimum design requirements for combustion control system

The combustion control system shall meet operational requirements and correctly interface with the process. To accomplish this objective, the following requirements are defined for minimum system design:

(1) Process measurement requirements

(2) Control and logic requirements

(3) Final control device requirements

(4) System reliability and availability

(5) Alarm requirements

(6) Operator interface

4.1 Process measurement requirements

4.1.1 Instrument installation for combustion control Process sensing devices should be installed as close as practical to the source of the measurement with consideration being given to excessive vibration, temperature, and access for periodic maintenance. Recommendations for the location of instrument and control equipment connections can be found in the American Boiler Manufacturer Association (ABMA) "Recommendations for Location of Instrument and Control Connections for the Operation and Control of Watertube Boilers."

Separate isolation valves and impulse lines should be run to each pressure-sensing device used for control.

4.1.2 Measurement and conditioning Filtering techniques used to condition process measurements shall not adversely affect stability or reduce control system response.

4.1.3 Process measurements

4.1.3.1 Mass air flow measurement The mass air flow measurement shall be a repeatable signal that is representative of the air entering the furnace.

When volumetric air flow rate measurement techniques are employed and the air temperature at the flow-measuring element varies 50°F (28°C) or more, the measured (indicated) flow shall be compensated for flowing air density to determine the true mass air flow rate.

4.1.3.2 Furnace pressure measurement Furnace pressure shall be measured with three furnace pressure transmitters, each on a separate pressure-sensing tap.

4.1.3.3 Fuel measurement The fuel flow measurement shall be a representative measure of the total fuel energy entering the furnace.

4.1.3.4 Gas analysis measurement A representative flue gas oxygen measurement shall be provided. A representative flue gas sample of equivalent combustibles should be provided.

4.2 Control and logic requirements

4.2.1 Automatic tracking shall be provided for bumpless control mode transfer.

4.2.2 The combustion control, which responds to the boiler energy demand, shall be accomplished with the following:

(1) Furnace pressure (balanced draft systems) control

(2) Air demand and air control

(3) Fuel demand and fuel control

The development of the boiler energy demand is covered in another SP77 fossil power plant standard.

4.2.3 Furnace pressure (draft) control The furnace pressure control shall regulate flue gas flow to maintain furnace pressure at the desired set point in compliance with the requirements of NFPA 85 G.

The furnace pressure control shall utilize a feedforward demand signal from the air flow control.

4.2.4 Air demand and air control Air flow demand shall be developed from the boiler energy demand and used to control air flow to the furnace.

There shall be a minimum air flow demand limit to prevent air from being reduced below the level required to support stable flame conditions in the furnace when the air control is in automatic. Suitable provisions shall be included to prevent controller windup under minimum air limit conditions.

The minimum air flow limit shall be in compliance with the requirements of the NFPA 85 series.

The following are prerequisites for the air flow control in automatic:

(1) Furnace pressure control in automatic (balanced draft systems).

(2) One forced draft fan (or other air source) in service and the associated regulating device (s) in automatic control.

Provision shall be made to ensure that the automatic regulation of air shall result in a fuel-to-air ratio that provides safe boiler operation. This shall include limiting of fuel flow or air flow to ensure that fuel flow never exceeds the safe combustion limit that the air flow will support.

4.2.5 Fuel demand and fuel control Fuel demand shall be developed from the boiler energy demand and used to control fuel flow to the furnace.

Total fuel input shall be determined from one or a combination of calculated values, fuel measurements, or characterized fuel demand outputs.

The fuel demand/fuel input relationship shall be used to control energy balance on a Btu (kJ) basis. When the fuel control is in automatic, there shall be a minimum fuel demand limit to prevent fuel from being reduced below the level required to support stable flame conditions in the furnace. Suitable provisions shall be included to prevent controller windup under minimum fuel limit conditions.

The following are prerequisites for fuel controller in automatic:

(1) Air control in automatic

(2) One fuel source in service and the associated regulating device (s) in automatic control

Provision shall be made to ensure that the automatic regulation of fuel shall result in a fuel-to-air ratio that provides safe boiler operation. This shall include limiting of fuel flow or air flow under all conditions to ensure that fuel flow never exceeds the safe combustion limit that the air flow will support.

4.2.6 Excess air Excess air shall be maintained at all loads to assure proper combustion of the fuel entering the furnace and should not allow the furnace to operate at an oxygen level in the flue gas below the boiler or burner manufacturer's requirements.

4.3 Final control device requirements All final control elements shall be designed to fail safe on loss of demand signal or motive power, i.e., open, close, or lock in place. The fail-safe position shall be determined by the user and be based upon the specific application.

4.4 System reliability and availability In order to establish minimum criteria, the combustion control system specification shall include the following as part of the system design base:

(1) Maximum and minimum unit load (steaming capacity)

(2) Normal operating load range

(3) Anticipated load changes

(4) Start-up and shutdown frequency

(5) Degree of automation

(6) Boiler auxiliary maximum and minimum capacities

Field transmitting device redundancy should be provided to the extent necessary to achieve desired system reliability.

When two transmitters are employed, excessive deviation between the transmitters shall be alarmed, and the associated control loop shall be transferred to manual.

When three transmitters are employed, excessive deviation between the transmitters shall be alarmed. A transmitter select scheme shall be used for control purposes.

4.5 Minimum alarm requirements

Minimum alarm requirements shall include the following:

4.5.1 Process alarms

(1) Low air flow

(2) Low flue gas oxygen

(3) High/low furnace pressure

(4) Fuel alarms per NFPA 85 series

4.5.2 Hardware alarms

(1) Loss of control power

(2) Loss of final drive power

(3) Control loop transfer to manual due to hardware failure

(4) Failure of process measurement signal

(5) Redundant transmitters deviation alarm

(6) Final drives at control limit

(7) Loss of redundant components

4.6 Operator interface

4.6.1 The following information used in the combustion control system shall be made available to the operator:

(1) Boiler energy demand

(2) Air flow

(3) Total fuel flow (measured or inferred)

(4) Individual fuel flow (s) (according to fuel type for multiple fuels)

(5) Furnace pressure

(6) Flue gas oxygen content

(7) All alarms previously defined in 4.5

(8) Control loop status and output demand

(9) Set points

(10) Bias

4.6.2 In addition to the above, the following information, as applicable, should be made available to the operator:

(1) Final drive position (s) and demand deviation status

(2) Valve position (s)

(3) Feeder speeds

(4) Control power status

(5) Drive motive power status

(6) Oil or gas burner pressure

4.6.3 The system shall include capabilities for the automatic and/or manual control of each individual final control device.

References

1) ANSI/IEEE Std. 268-1982, Metric Practice, American National Standards Institute, 11 West 42nd Street, New York, NY 10036.

2) ANSI/ISA-S5.1-1984, Instrumentation Symbols and Identifications, Instrument Society of America, 67 Alexander Drive, P. O. Box 12277, Research Triangle Park, N.C. 27709.

3) ANSI/ISA-S5.4-1976 (R 1981), Instrument Loop Diagrams, Instrument Society of America.

4) ANSI/ISA-S51.1-1979, Process Instrumentation Terminology, Instrument Society of America.

5) ASME Boiler and Pressure Vessel Code, Section 1, American Society of Mechanical Engineers, 345 East 437th Street, New York, NY 10017, 1983.

6) API RP550 Pt. IV-1984. Manual on Installation of Refinery Instruments and Control Systems: Steam Generators, American Petroleum Institute, 1220 L Street, NW, Washington, DC 20005.

7) ASTM Annual Book of Gaseous Fuels: Coal and Coke Atmospheric, Parts 19 and 26, American Society for Testing Material.

8) Babcock & Wilcox Steam - Its Generation and Use, 39th ed., New York, NY, 1978.

9) IEEE Std. 502-1985, "IEEE Guide for Protection, Interlocking and Control of Fossil-Fueled Unit Connection Steam Stations," Institute of Electrical and Electronics Engineers, 345 East 47th Street, New York, NY 10017, 1985.

10) NFPA 85 Pamphlet Series, National Fire Protection Association, Quincy, MA 02269, 1988 Editions. These are as follows: NFPA - 85 A (1987); 85 B (1989); 85 C (1991); 85 D (1989); 85 E (1985); 85 F (1988); 85 G (1987).

11) PMC 20.1-1973, Process Measurement and Control Terminology, Scientific Apparatus Makers Assoc., 1101 16th Street, NW, Ste.300, Washington, DC 20036.

12) PMC 22.1-1981, Functional Diagramming of Instrument and Control Systems, Scientific Apparatus Makers Association.

13) Singer, Joseph G., Combustion: Fossil Power Systems, 3rd ed., Combustion Engineering, Inc., 1981.

14) Skretzki, Bernhardt G.A., "Electric Generation–Steam Stations."

15) UltraSystems, Inc. (for Naval Civil Engineering Laboratory), "Theory of Boiler Control Systems - Operation Manual," N62474-81-C-9388, Irvine, CA, 142 pp.

16) SAMA/ABMA/IGCI's Std., Instrument Connections Manual, "Recommendations for Location of Instrument and Control Connections for the Operation and Control of Watertube Boilers," Scientific Apparatus Makers Association and Industrial Gas Cleaning Institute, Inc., ABMA, 950 North Glebe Road, Suite 180, Arlington, VA 22203; IGCI, 700 North Fairfax Street, Suite 304, Alexandria, VA 22314, 1981.

Appendix A

General tutorial

This appendix is included for information purposes and, therefore, is not considered a part of the ISA-S77.41.

A.1 Purpose

The purpose of this appendix is to provide tutorial information on the philosophy underlying this standard and to assist in specifying and applying combustion control strategies that will best serve the requirements of the user.

A.2 Introduction

The purpose of any combustion control system is to safely and efficiently maintain the desired boiler output without the need for constant operator attention. Therefore, the combustion process inside the furnace must be controlled while the boiler output changes in response to load demands. The basic principle of combustion control is to meet the boiler load requirements by regulating the quantities of fuel and air while achieving optimum combustion and maintaining safe conditions for operators and equipment.

A.3 Combustion process

As the combustion process takes place in the furnace, oxygen in the combustion air combines chemically with the carbon and hydrogen in the fuel to produce heat. The amount of air that contains enough oxygen to combine with all the combustible matter in the fuel is called the "stoichiometric" value or theoretical air.

It is improbable for every molecule of fuel that enters the furnace to combine chemically with oxygen. For this reason, it is necessary to provide more air than the stoichiometric requirement. For most boilers it is customary to provide 5 to 20 percent more air than the stoichiometric requirement to ensure complete combustion. This additional air is called "excess air." A boiler firing at 1.2 times the stoichiometric air requirement would be said to be firing at 20 percent excess air.

If insufficient oxygen is introduced into the furnace, incomplete combustion of the fuel will occur. This wastes fuel, causes air pollution, and results in hazardous conditions in the boiler. The unburned fuel may ignite in the boiler or breeching and result in secondary combustion, causing a dangerous explosion.

Providing too much combustion air reduces the explosion danger but also reduces efficiency. The largest energy loss in the boiler is the heat that escapes as hot flue gas. Increasing the excess air increases this energy loss. High excess air can also result in unstable burner conditions due to the lean fuel/air mixture.

In practice, a large number of items that affect boiler efficiency are related to excess air. The proper value of excess air is a function of boiler load, fuel quantity, air leakage through idle burners, steam temperature, flame stability, and energy losses.

A.4 Basic combustion control strategies

Various combustion control strategies that are based on requirements for safe, efficient, and responsive control of boilers have evolved. More sophisticated controls were developed as instrumentation became more reliable and accurate. The control strategies can be divided into two major categories: positioning systems and metering systems.

In positioning systems the fuel and air control devices are simultaneously positioned, based on energy demand. Each position of the fuel control device assumes a corresponding position for the airflow control device. A control station is normally available for the operator to trim the fuel/air ratio.

The positioning system is simple and fast responding, but it cannot compensate for varying fuel characteristics, atmospheric conditions, dynamic characteristics of the fuel delivery equipment, or the imbalance of the fuel-to-air ratio during rapid load changes.

Metering systems measure the actual fuel and air delivered to the boiler. The measured flows are used in feedback control schemes to precisely regulate the fuel-to-air ratio. Air flows can be measured without too much difficulty. The fuel flow in a gas or oil boiler can also be readily measured. Fuel flow in a coal-fired boiler cannot be directly measured, and various schemes have been developed to infer the fuel delivery rate based on other variables. In addition, the heavy equipment necessary to transport the coal and prepare it for burning present dynamic operational and control problems.

A typical combustion control functional diagram is shown in Figure A.1.

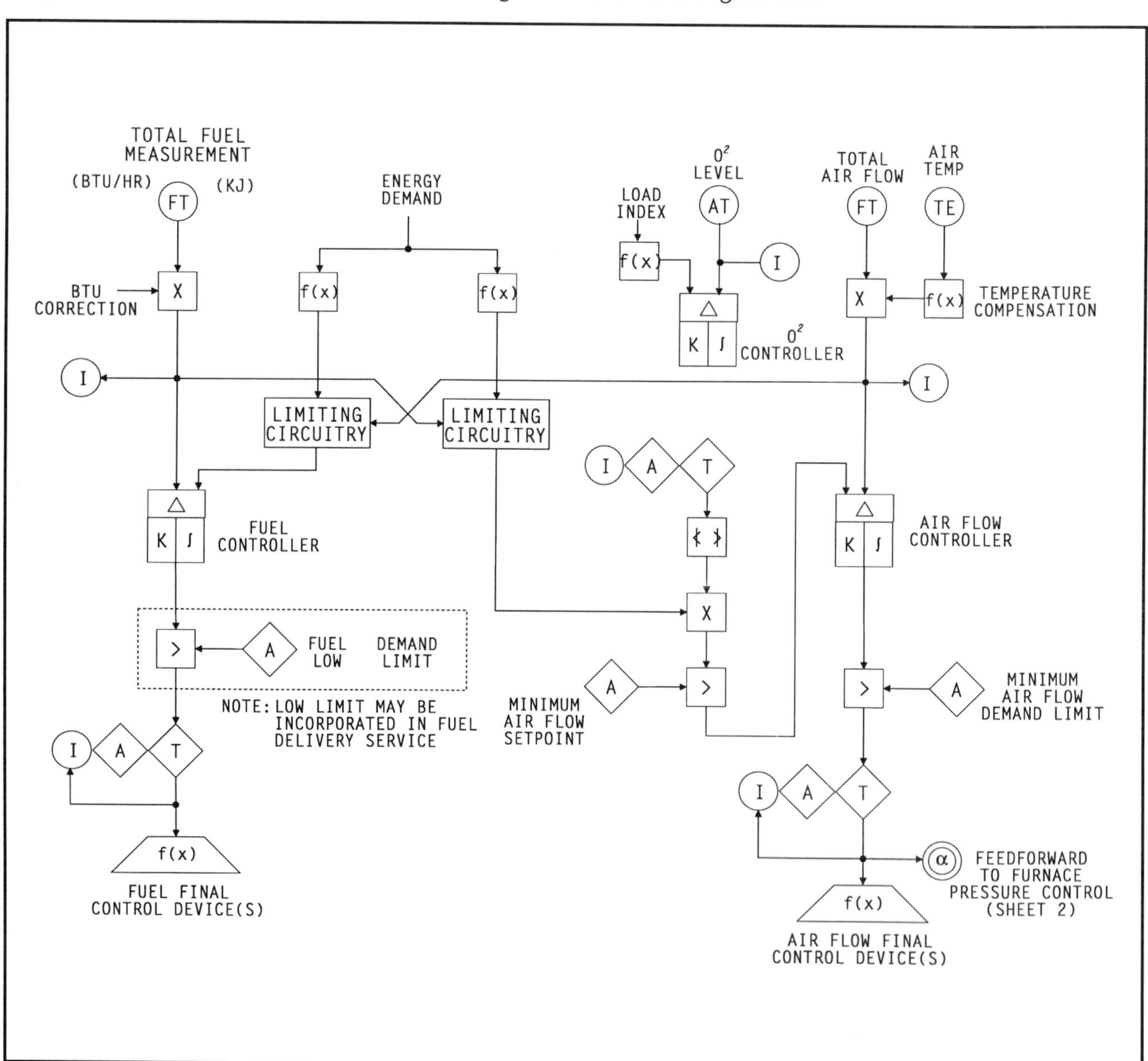

Figure A.1. Typical combustion control functional diagram.

A.5 Air flow requirements concepts

Air flow measurements can be established by primary elements located on either the air side or the flue-gas side of the unit.

A suitable place to measure air flow is the entry of the forced draft fan. When there is potential leakage between the forced draft fan and the furnace, a preferred location may be the air duct between the forced draft fan and the burners downstream of the air heater.

An alternate, but less preferred, location for air flow measurement is on the flue-gas side of the boiler. Connections for a differential pressure transmitter are located across a section of the boiler, and the pressure drop of the flue gas is measured. The connections should be located to avoid the pressure drop across the air heater. Potential problems with this method are the corrosive effects of the flue gas and leakage.

A.6 Furnace pressure (draft)

Furnace pressure (draft) control is required on balanced draft boilers. While either the forced draft fan (s) or the induced draft fan (s) could be used to control the furnace pressure, typically the induced draft fan (s) are used. A typical furnace pressure control functional diagram is shown in Figure A.2.. It utilizes a feedforward signal characterized to represent the position of the forced draft control device (s). In a properly designed and calibrated system, the output of the furnace pressure controller will remain near its midrange for all air flows.

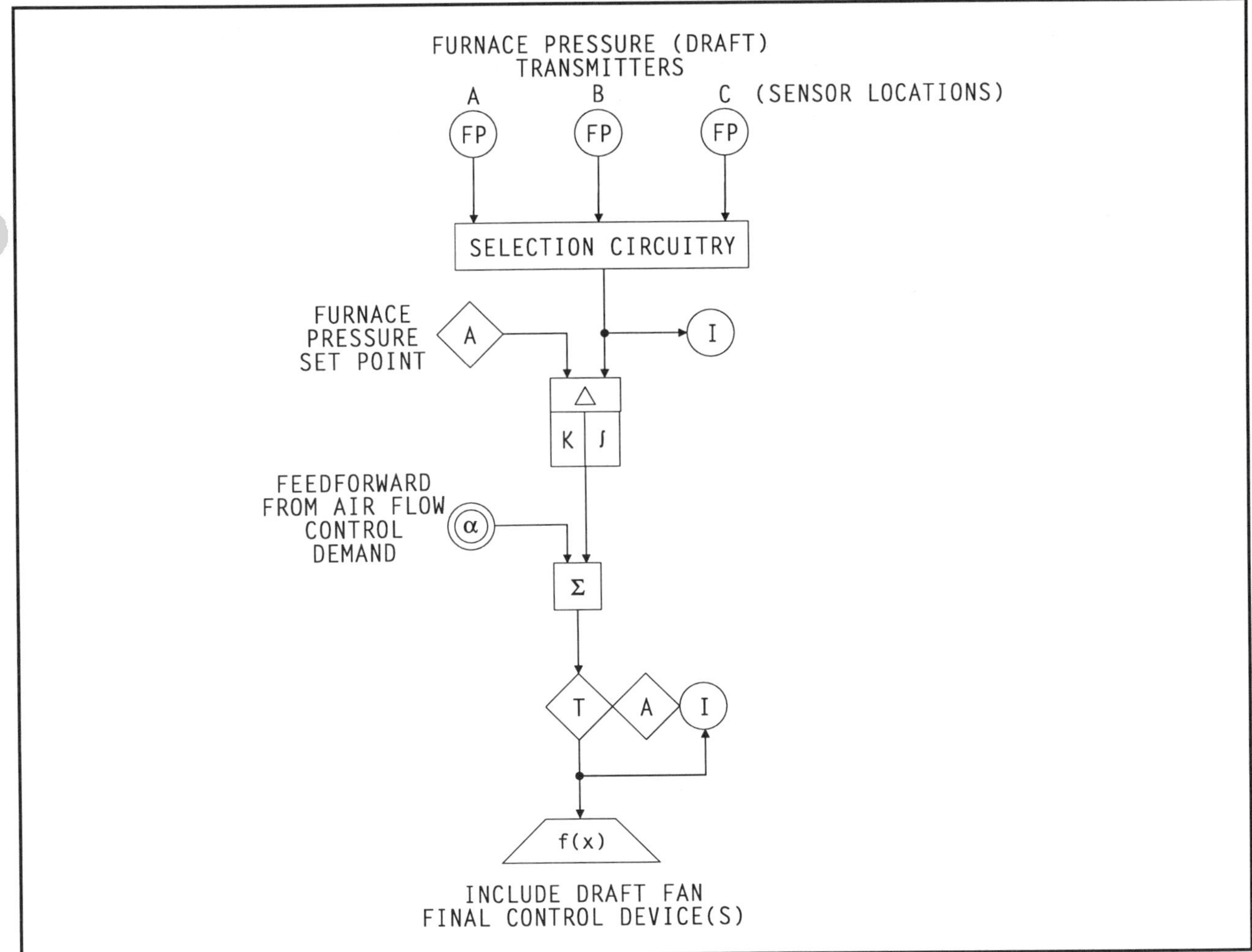

Figure A.2. Typical furnace pressure control functional diagram.

A.7 Control of gas and oil fired boilers

Gas - and oil-fired boilers often allow burning gas and oil separately or together. Provisions must be made to assure that the fuel-to-air ratio is maintained as multiple fuels are introduced into the boiler. This can be accomplished by summing the individual fuel flows on a Btu (kJ) value basis. The output of the summer becomes the "total fuel" in the calculation of the fuel-to-air ratio. Limiting circuitry should be provided to ensure a safe fuel-to-air ratio.

A.8 Control of pulverized coal-fired boilers

Two major considerations in the development of a metering-type control strategy for a pulverized coal-fired boiler are compensation for pulverizer dynamics and the selection of a inferred coal flow signal. In addition, the fuel demand signal must be corrected for the number of pulverizers in service.

A.8.1 Pulverizer dynamics

Pulverizers induce time lags into the control process. There is a finite time between the introduction of coal into the pulverizer and the delivery of fuel to the furnace. The grinding and drying of the coal in the pulverizer and the time required to transport the coal/air mixture to the burners both contribute to time lags. In addition, the volume of coal in the pulverizer increases as the load on the pulverizer increases.

A.8.2 Ball tube pulverizers

The ball tube or Hardinge-type pulverizers utilize small diameter (2 inch) steel balls in a rotating drum to grind coal. These pulverizers exhibit large coal storage characteristics. It is common for this type of pulverizer to produce controlled coal flow for 10 to 15 minutes after the coal feeder is stopped. The coal feeder signal is, therefore, useless in the control scheme as a fuel flow signal. An implied fuel flow signal can be derived from boiler heat release, primary air flow, linearized classifier differential, characterized primary air damper position, or exhauster inlet damper position. Since these signals, with the exception of heat release, can produce an apparent coal flow in an empty mill, the signals must be verified by logic signals monitoring the mill level and feeder status.

In a ball tube pulverizer, the coal feeder normally operates independently of the boiler energy demand or fuel demand signals. The coal feeder control for ball tube pulverizers may be based on the following:

1) Pulverizer level (differential measurement)
2) Pulverizer dB (sound measurement)
3) Pulverizer kW (power measurement)
4) Pulverizer kW and dB

The usual control of the feeder is based on pulverizer coal level with a slight feedforward or derivative from boiler energy demand or fuel demand.

A8.3 Non-ball tube mills

Pulverized coal-fired boilers normally use a sophisticated combustion control system having several pulverizers, each supplying multiple burners. It is not only important to maintain the correct fuel-to-air ratio at all times, but the fuel from each pulverizer to its associated burner should be properly proportioned and distributed for stable and efficient boiler operation.

The firing rate demand is compared to the total measured fuel flow (summation of all feeders in service delivering coal) to develop the demand to the fuel controller. The fuel demand signal is then applied in parallel to all operating pulverizers.

Should an upset in available air occur so that air is limited, an error signal from air flow control should reduce the firing rate demand to the fuel controller to maintain a minimum acceptable fuel/air ratio. Limiting circuitry shall be provided to ensure that air flow is always above demanded fuel flow; hence, a safe combustion mixture is always present. This demand is compared with total feeder speeds or the heat absorbed signal. Should there be a difference between the fuel demand and total fuel flow, the fuel controller will readjust the speed of the feeders in service to the extent necessary to eliminate the error.

Since there may be some delay between a change in feeder speed and the actual change in coal to a furnace, a lag (mill model) is incorporated into the speed feedback, so that the air flow and fuel flow control are kept in step with the actual coal to the furnace. During conditions of a mill overload, the feeder speed demand signal should be reduced until the overload condition is resolved.

A.8.4 Reference for pulverizer control

The pulverizer manufacturer is the prime source for information concerning recommended control strategies.

Appendix B

Fault tolerance

This appendix is included for information purposes and, therefore, is not considered a part of ISA-S77.41.

The ability of a combustion control system to tolerate failures and still provide an acceptable level of performance is influenced by the user's decisions regarding system design. Typically, the vendor's hardware is modular and can be configured in many forms. The way the system design utilizes these modules has a significant effect on the ability of overall systems to tolerate failures. A careful review of the entire system is necessary to determine the effect of equipment failures and the ability of the operator to continue operations.

B.1 Geographic distribution

Geographic distribution results from placing the hardware (controllers, data acquisition modules, power supplies, etc.) in physically remote areas of the plant and connecting them via a digital communications network. This hardware is normally placed in a convenient location close to the equipment being monitored or controlled.

While geographic distribution may enhance reliability, it will also require additional effort and expense if redundant power sources, air conditioned environments, or redundant communication channels are necessary. Often, geographic distribution is used to reduce wiring costs or relieve overcrowded equipment rooms.

B.2 Functional distribution

The goal of functional distribution is to limit the effects of a hardware failure to an acceptable level. By partitioning the system into modules, the effect of a single hardware failure is minimized to predefined situations. All of the elements of a specific control task (i.e., furnace pressure, pulverizer A, etc.) are segregated into a module capable of independent operation. Each module should be capable of continued operation at some minimal level without depending on inputs from other modules. Aspects to be considered are the power source (s), transmitters, operator interface, input/output circuits, communications links, and computational requirements.

B.3 Fail-safe positioning

Final drive units must be evaluated to determine the preferred failure mode upon loss of motive power. Most drive units can be selected to fail open, fail closed, or fail in place. In addition, the effect of the loss of the driving signal must be considered. In some cases, air-to-close (or increasing current-to-close) dampers are preferred.

B.4 Redundant component

The modular design of most vendors' equipment allows the user to take a building block approach to selecting components. Modules can be added or deleted as needed. In most cases, the number of inputs, outputs, and computational power are flexible within certain limits. This allows the user to add redundant modules to achieve the desired level of fault tolerance.

B.4.1 Power supplies

Power supplies driven from separate primary feeds should be considered, especially for operator interface consoles. Each power supply should be capable of continuous operation in the event of failure of the other. Battery backup is also an option.

B.4.2 Communications

As a minimum, the communication channels should be redundant. If portions of the system are geographically distributed, separate cable routes should be capable of continuous operation in the event of failure of the other.

B.4.3 Controllers

Redundant controllers can be utilized in a 1-to-1 configuration or in a 1-to-*N* configuration. In the 1-to-1 configuration, each controller has a backup controller. In the 1-to-*N* configuration, one backup controller monitors the operation of several controllers and automatically replaces the functions of a failed unit but cannot support multiple controller failures.

B.4.4 Operator interface

Console units, manual/automatic stations, and failure modes must be considered together when evaluating operator interfaces. No firm rules exist for minimum requirements concerning redundancy. The goal should be to maintain the operator's ability to control the boiler in spite of any equipment failures. Multiple, functionally independent console units may be preferred to a single console with manual/automatic stations serving as backup.

If multiple console units are used to achieve redundance, the consoles should be as independent as possible with due regard to economics and convenience. As a guideline, a single equipment failure or loss of a power source should not prevent the operator from responding to emergency situations.

Manual/automatic stations have traditionally been considered "fail-safe" components. However, manual/automatic stations may incorporate a significant number of electronic components that must remain functional for the operator to have "hard manual" control of the process.

Strong consideration should be given to the use of actual final element position through the position feedback sensors for display on M/A stations and for use in the operation of the combustion control system in lieu of "demand output."

B.4.5 Field devices

Field device redundancy should be provided to the extent necessary to achieve the required reliability and meet the requirements of the standard. In most cases, this requires a case-by-case consideration of critical parameters.

ISA STANDARDS AND RECOMMENDED PRACTICES

RP2.1 — Manometer Tables
1978, 31 pp.

S5.1 — Instrumentation Symbols and Identification (Formerly ANSI Y32.20)
ANSI/ISA-1984
1984, 64 pp.

S5.2 — Binary Logic Diagrams for Process Operations
ANSI/ISA-1976 (R 1981)
Reaffirmed-1981, 19 pp.

S5.3 — Graphic Symbols for Distributed Control/Shared Display Instrumentation, Logic and Computer Systems
1982, 16 pp.

S5.4 — Instrument Loop Diagrams
ANSI/ISA-1991
1991, 18 pp.

S5.5 — Graphic Symbols for Process Displays
ANSI/ISA-1985
1986, 44 pp.

RP7.1 — Pneumatic Control Circuit Pressure Test
1956, 6 pp.

S7.3 — Quality Standard for Instrument Air
ANSI/ISA-1975 (R 1981)
Reaffirmed-1981, 6 pp.

S7.4 — Air Pressures for Pneumatic Controllers, Transmitters, and Transmission Systems
ANSI/ISA-1981
1981, 6 pp.

RP7.7 — Recommended Practice for Producing Quality Instrument Air
1984, 16 pp.

S12.1 — Definitions and Information Pertaining to Electrical Instruments in Hazardous (Classified) Atmospheres
1991, 64 pp.

S12.4 — Instrument Purging for Reduction of Hazardous Area Classification
1970, 12 pp.

RP12.6 — Installation of Intrinsically Safe Systems for Hazardous (Classified) Locations
ANSI/ISA-1987
1987, 12 pp.

S12.10 — Area Classification in Hazardous (Classified) Dust Locations
ANSI/ISA-1988
1988, 29 pp.

S12.12 —Electrical Equipment for Use in Class I, Division 2 Hazardous (Classified) Locations
ANSI/ISA-1984
1984, 26 pp.

S12.13, Part I — Performance Requirements Combustible Gas Detectors
ANSI/ISA-1986
1986, 24 pp.

RP12.13, Part II — Installation, Operation, and Maintenance of Combustible Gas Detection Instruments
ANSI/ISA-1987
1987, 156 pp.

S12.15, Part I — Performance Requirements for Hydrogen Sulfide Detection Instruments (10-100 ppm)
1990, 28 pp.

RP12.15 Part II — Installation, Operation, and Maintenance of Hydrogen Sulfide Detection Instruments
1990, 29 pp.

RP16.1,2,3 — Terminology, Dimensions and Safety Practices for Indicating Variable Area Meters (Rotameters, Glass Tube, Metal Tube, Extension Type Glass Tube)
1959, 6 pp.

RP16.4 — Nomenclature and Terminology for Extension Type Variable Area Meters (Rotameters)
1960, 3 pp.

RP16.5 — Installation, Operation, Maintenance Instructions for Glass Tube Variable Area Meters (Rotameters)
1961, 6 pp.

RP16.6 — Methods and Equipment for Calibration of Variable Area Meters (Rotameters)
1961, 7 pp.

S18.1 — Annunciator Sequences and Specifications
ANSI/ISA-1979 (R 1992)
1992, 36 pp.

S20 — Specification Forms for Process Measurement and Control Instruments, Primary Elements and Control Valves
Reaffirmed-1981, 72 pp.

S26 — Dynamic Response Testing of Process Control Instrumentation
1975, 25 pp.

RP31.1 — Specification, Installation, and Calibration of Turbine Flowmeters
ANSI/ISA-1977
1977, 21 pp.

RP37.2 — Guide for Specifications and Tests for Piezoelectric Acceleration Transducers for Aerospace Testing
Reaffirmed-1982, 19 pp.

RP42.1 — Nomenclature for Instrument Tube Fittings
1992, 13 pp.

S50.1 — Compatibility of Analog Signals for Electronic Industrial Process Instruments
ANSI/ISA-1982
Reaffirmed-1991, 11 pp.

S50.02, Part 2 — Fieldbus Standard for Use in Industrial Control Systems – Part 2: Physical Layer Specification and Service Definition
ISA-1992
1992, 115 pp.

S51.1 — Process Instrumentation Terminology
ANSI/ISA-1979
1979, 41 pp.

RP52.1 — Recommended Environments for Standards Laboratories
1975, 18 pp.

RP55.1 — Hardware Testing of Digital Process Computers
ANSI/ISA-1975 (R 1983)
Reaffirmed-1983, 54 pp.

RP60.1 — Control Center Facilities
1990, 16 pp.

RP60.3 — Human Engineering for Control Centers
1990, 17 pp.

RP60.4 — Documentation for Control Centers
1990, 22 pp.

RP60.6 — Nameplates, Labels and Tags for Control Centers
1984, 35 pp.

RP60.8 — Electrical Guide for Control Centers
1978, 6 pp.

RP60.9 — Piping Guide for Control Centers
1981, 12 pp.

RP60.11 — Crating, Shipping and Handling for Control Centers
1991, 35 pp.

S61.1 — Industrial Computer System FORTRAN Procedures for Executive Functions, Process Input-Output, and Bit Manipulation
ANSI/ISA-1977
1976, 11 pp.

S61.2 — Industrial Computer System FORTRAN Procedures for File Access and the Control of File Contention
ANSI/ISA-1978
1978, 7 pp.

S67.01 — Transducer and Transmitter Installation for Nuclear Safety Applications
ANSI/ISA-1979 (R 1987)
1987, 16 pp.

S67.02 — Nuclear Safety-Related Instrument Sensing Line Piping and Tubing Standards for Use in Nuclear Power Plants
ANSI/ISA-1983
1983, 32 pp.

S67.03 — Standard for Light Water Reactor Coolant Pressure Boundary Leak Detection
ANSI/ISA-1982
1982, 28 pp.

S67.04 — Setpoints for Nuclear Safety-Related Instrumentation
ANSI/ISA-1988
1987, 13 pp.

S67.06 — Response Time Testing of Nuclear Safety-Related Instrument Channels in Nuclear Power Plants
ANSI/ISA-1984
1984, 17 pp.

S67.10 — Sample-Line Piping and Tubing Standard for Use in Nuclear Power Plants
ANSI/ISA-1986
1986, 22 pp.

S67.14 — Qualifications and Certification of Instrumentation and Control Technicians in Nuclear Power Plants
ANSI/ISA-1983
1983, 16 pp.

S71.01 — Environmental Conditions for Process Measurement and Control Systems: Temperature and Humidity
ANSI/ISA-1986
1985, 20 pp.

S71.02 — Environmental Conditions for Process Measurement and Control Systems: Power
1991, 13 pp.

S71.04 — Environmental Conditions for Process Measurement and Control Systems: Airborne Contaminants
ANSI/ISA-1986
1986, 16 pp.

ISA STANDARDS AND RECOMMENDED PRACTICES

S72.01 — PROWAY-LAN Industrial Data Highway
ANSI/ISA-1985
1985, 186 pp.

RP74.01 — Application and Installation of Continuous-Belt Weighbridge Scales
1984, 20 pp.

S75.01 — Flow Equations for Sizing Control Valves (R 1985)
ANSI/ISA-1985
1985, 28 pp.

S75.02 — Control Valve Capacity Test Procedure
ANSI/ISA-1982
1988, 20 pp.

S75.03 — Face-to-Face Dimensions for Integral Flanged Globe-Style Control Valve Bodies (ANSI) Classes 125, 150, 250, 300, 600)
ISA-1992
1992, 12 pp.

S75.04 — Face-to-Face Dimensions for Flangeless Control Valves
(ANSI Classes 150, 300, 600)
ANSI/ISA-1985 (R 1992)
Reaffirmed 1992, 10 pp.

S75.05 — Control Valve Terminology
ANSI/ISA-1983
1983, 33 pp.

RP75.06 — Control Valve Manifold Designs
1981, 20 pp.

S75.07 — Laboratory Measurement of Aerodynamic Noise Generated by Control Valves
ANSI/ISA-1987
1987, 16 pp.

S75.08 — Installed Face-to-Face Dimensions for Flanged Clamp or Pinch Valves
ANSI/ISA-1985
1985, 14 pp.

S75.11 — Inherent Flow Characteristic and Rangeability of Control Valves
ANSI/ISA-1985 (R 1991)
1991, 10 pp.

S75.12 — Face-to-Face Dimensions for Socket Weld-End and Screwed-End Globe-Style Control Valves (ANSI Classes 150, 300, 600, 900, 1500, and 2500)
1993, 13 pp.

S75.13 — Method of Evaluating Performance of Positioners with Analog Input Signals and Pneumatic Output
1989, 29 pp.

S75.14 — Face-to-Face Dimensions for Buttweld-End Globe-Style Control Valves (ANSI Class 4500)
1993, 11 pp.

S75.15 — Face-to-Face Dimensions for Buttweld-End Globe-Style Control Valves (ANSI Classes 150, 300, 600, 900, 1500, and 2500)
1993, 13 pp.

S75.16 — Face-to-Face Dimensions for Flanged Globe-Style Control Valve Bodies (ANSI Classes 900, 1500, and 2500)
1993, 13 pp.

S75.17 — Control Valve Aerodynamic Noise Prediction
ANSI/ISA-1991
1991, 20 pp.

RP75.18 — Control Valve Position Stability
1989, 14 pp.

S75.19 — Hydrostatic Testing of Control
1989, 22 pp.

S75.20 — Face-to-Face Dimensions for Separable Flanged Globe-Style Control Valves
1991, 12 pp.

RP75.21 — Process Data Presentation for Control Valves
1989, 17 pp.

S75.22 — Face-to-Centerline Dimensions for Flanged Globe-Style Angle Control Valve Bodies
(ANSI classes 150, 300, and 600)
1992, 8 pp.

S77.41 — Fossil Fuel Power Plant Boiler Combustion Controls
1992, 19 pp.

S77.42 — Fossil-Fuel Power Plant Feedwater Control System—Drum-Type
ANSI/ISA-1987
1987, 24 pp.

S82.01 — Safety Standard for Electrical and Electronic Test, Measuring, Controlling and Related Equipment—General Requirements
ANSI/ISA-1988
1988, 65 pp.

S82.02 — Safety Standard for Electrical and Electronic Test, Measuring, Controlling and Related Equipment—Electrical and Electronic Test and Measuring Equipment
ANSI/ISA-1988
1988, 18 pp.

S82.03 — Safety Standard for Electrical and Electronic Test, Measuring, Controlling and Related Equipment—Electrical and Electronic Process Measurement and Control Equipment
ANSI/ISA-1988
1988, 19 pp.

MC96.1 — American National Standard for Temperature Measurement Thermocouples
1982, 48 pp.

ANSI C100.6-3 — American National Standard for Voltage or Current Reference Devices; Solid State Devices
1984, 14 pp.